THE FIRST SYSTEMS OF

WEIGHTED

DIFFERENTIAL AND INTEGRAL

CALCULUS

*

Jane Grossman
University of Lowell

Michael Grossman
University of Lowell

Robert Katz
Archimedes Foundation

Archimedes Foundation
Box 240, Rockport
Massachusetts 01966

ISBN 0977117014

"In general the position as regards all such new calculi is this - That one cannot accomplish by them anything that could not be accomplished without them. However, the advantage is, that, provided such a calculus corresponds to the inmost nature of frequent needs, anyone who masters it thoroughly is able - without the unconscious inspiration of genius which no one can command - to solve the respective problems, yea, to solve them mechanically in complicated cases in which, without such aid, even genius becomes powerless. Such is the case with the invention of general algebra, with the differential calculus, and in a more limited region with Lagrange's calculus of variations, with my calculus of congruences, and with Möbius's calculus. Such conceptions unite, as it were, into an organic whole countless problems which otherwise would remain isolated and require for their separate solution more or less application of inventive genius."

Gauss

First Printing, July 1980

P R E F A C E

Mathematicians have long known about weighted averages, Stieltjes integrals, and derivatives of one function with respect to another function. But apparently no one has hitherto noticed that by means of suitably modified definitions those three concepts can be packaged into systems of calculi, complete with two fundamental theorems, mean value theorems, and analogues of other important theorems of classical calculus. We refer to the new systems as weighted calculi, because in each such system a weight function plays a central role.

In Chapter 1 we present the classical calculus in a novel manner that leads naturally to the subsequent construction of the weighted calculi.

Chapter 2 contains a development of the weighted classical calculi, which involve weighted arithmetic averages, weighted classical integrals (these are Stieltjes integrals), and weighted classical derivatives (these are classical derivatives of one function with respect to another function). We were surprised to discover that many of the weighted classical calculi are (unweighted) non-Newtonian calculi. (The non-Newtonian calculi were discovered by Michael Grossman and Robert Katz and discussed in their book *Non-Newtonian Calculus*, which was published by Lee Press in 1972. A familiarity with that book is presupposed only for Section 2.11 and Chapter 3.)

Chapter 3 includes a development of the weighted non-Newtonian calculi, which involve weighted non-Newtonian averages, weighted non-Newtonian integrals, and weighted non-Newtonian derivatives. The chapter concludes with a brief presentation of weighted geometric calculus, an interesting example of a weighted non-Newtonian calculus.

Since this book is intended for a wide audience, including students, engineers, scientists, as well as mathematicians, we have presented many details that would not appear in a research report and we have excluded proofs. (Most of the results can be proved in a straightforward way.) It is assumed, of course, that the reader has a working knowledge of the rudiments of classical calculus.

Jane Grossman
Michael Grossman
Robert Katz

July 14, 1980

C O N T E N T S

PRELIMINARIES

The word number means real number. The letter R stands for the set of all numbers.

If $r < s$, then the interval $[r, s]$ is the set of all numbers x such that $r \leqq x \leqq s$. (Only such intervals are used here.) The interior of $[r, s]$ consists of all numbers x such that $r < x < s$. The classical extent of $[r, s]$ is $s - r$. A unit interval is any interval whose classical extent is 1.

An arithmetic partition of $[r, s]$ is any arithmetic progression whose first term is r and last term is s. An arithmetic partition with exactly n terms is said to be n-fold.

A point is any ordered pair of numbers, each of which is called a coordinate of the point. A function is any set of points, each distinct two of which have distinct first coordinates.

The domain of a function is the set of all its arguments (first coordinates); the range of a function is the set of all its values (second coordinates). A function is said to be on its domain, to be onto its range, and to be defined at each of its arguments.

The classical change of a function f on $[r, s]$ is $f(s) - f(r)$.

If every two distinct points of a function f have distinct second coordinates, then f is one-to-one and its inverse, f^{-1}, is the one-to-one function consisting of all points (y, x) for which (x, y) is a point of f.

A positive function is any function whose values are all positive; a discrete function is any function that has only a finite number of arguments.

The function exp is on R and assigns to each number x the number e^x, where e is the base of the natural logarithm function, ln. The function ln is the inverse of exp.

"for each successive class of phenomena, a new calculus or a new geometry, as the case might be, which might prove not wholly inadequate to the subtlety of nature."

Quoted, without citation,
by H. J. S. Smith; *Nature*,
Volume 8 (1873), page 450.

CHAPTER 1

Classical Calculus

1.0 INTRODUCTION

In this chapter we present the basic ideas of classical calculus in a novel manner that leads naturally to the subsequent construction of the systems of weighted calculus.

Although the term "classical analysis" is often used, the term "classical calculus" has appeared rarely, presumably because only one system of calculus was known prior to the discovery of the non-Newtonian calculi in 1967.

1.1 ARITHMETIC AVERAGE

The definition of the arithmetic average of a continuous function on an interval is based on the arithmetic average of n numbers, which is defined next.

The arithmetic average of n numbers $y_1, \ldots, y_n$ is the number $(y_1 + \cdots + y_n) / n$.

Let f be any continuous function on [r, s]. Then the arithmetic average of f on [r, s] is denoted by $M_r^s f$ and is defined to be the limit of the convergent sequence whose nth term is the arithmetic average

$$[f(a_1) + \cdots + f(a_n)] / n ,$$

where $a_1, \ldots, a_n$ is the n-fold arithmetic partition of [r, s].

The operator M is

<u>Additive</u>: $M_r^s(f + g) = M_r^s f + M_r^s g$,

<u>Homogeneous</u>: $M_r^s(c \cdot f) = c \cdot M_r^s f$, c any constant.

The operator M is characterized by the following three properties. (This use of the term "characterized" indicates that no other operator possesses all three properties.)

1. For any interval [r, s] and any constant function k(x) = c on [r, s],

$$M_r^s k = c \ .$$

2. For any interval [r, s] and any functions f and g continuous on [r, s], if $f(x) \leqq g(x)$ on [r, s], then

$$M_r^s f \leqq M_r^s g \ .$$

3. For any numbers r, s, t such that r < s < t and any function f continuous on [r, t],

$$(s - r) \cdot M_r^s f + (t - s) \cdot M_s^t f = (t - r) \cdot M_r^t f \ .$$

1.2 LINEAR FUNCTIONS

In classical calculus the linear functions are the standard to which other functions are compared.

A <u>linear function</u> is any function on R expressible in the form bx + c, where b and c are constants.

Each linear function u is 'uniform' in the following sense:

for any intervals $[r_1, s_1]$ and $[r_2, s_2]$ with the same classical extent,

$$u(s_1) - u(r_1) = u(s_2) - u(r_2);$$

that is, u has the same classical change on all intervals with the same classical extent.

Notice, in particular, that each linear function has the same classical change on all unit intervals.

In Section 1.4, we shall exploit the fact that there is exactly one linear function containing any two given points with distinct first coordinates.

It can easily be proved that if u is a linear function, then the ratio

$$\frac{u(s) - u(r)}{s - r}$$

has the same value on EVERY interval [r, s]. That fact suggests the definition of classical slope given next.

1.3 CLASSICAL SLOPE

The <u>classical slope of a linear function u</u> is the common value of all the ratios

$$\frac{u(s) - u(r)}{s - r},$$

where $r < s$. It is worth noting that the classical slope of u is equal to its classical change on any unit interval.

Of course, the classical slope of the linear function $bx + c$ turns out to be b.

1.4 CLASSICAL GRADIENT

The differential branch of classical calculus is rooted in the concept of the average rate of change, which we prefer to call the classical gradient. (The term "gradient" is better suited for our purpose, since it can readily be modified by appropriate adjectives. Of course, "gradient" is also used in vector analysis, but that subject does not concern us here.)

Let f be any function defined at r and s, where $r < s$. Then the <u>classical gradient of f on [r, s]</u> is denoted by $G_r^s f$ and is defined to be the classical slope of the linear function containing the points (r, f(r)) and (s, f(s)). It turns out that

$$G_r^s f = \frac{f(s) - f(r)}{s - r}.$$

If u is a linear function, then $G_r^s u$ equals the classical slope of u.

The operator G is

<u>Additive</u>: $G_r^s(f + g) = G_r^s f + G_r^s g$,

<u>Homogeneous</u>: $G_r^s(c \cdot f) = c \cdot G_r^s f$, c any constant.

Finally, when $r = s$, the expression for the classical gradient yields the indeterminate form 0/0, which brings us to the next topic.

1.5 CLASSICAL DERIVATIVE

Let f be any function defined at least on an interval containing the number a in its interior. If the following limit exists, we denote it by [Df](a) and call it the classical derivative of f at a:

$$\lim_{x \to a} \frac{f(x) - f(a)}{x - a} .$$

Notice that if a < x, then

$$\frac{f(x) - f(a)}{x - a} = G_a^x f .$$

And if x < a, then

$$\frac{f(x) - f(a)}{x - a} = G_x^a f .$$

The classical derivative of f, denoted by Df, is the function that assigns to each number t the number [Df](t), if it exists.

The classical derivative of a linear function has a constant value equal to its classical slope. Indeed, only linear functions have classical derivatives that are constant on R. In particular, if u is a function that is constant on R, then Du is everywhere equal to 0.

The operator D is

Additive: $D(f + g) = Df + Dg$,

Homogeneous: $D(c \cdot f) = c \cdot Df$, c any constant.

1.6 THE BASIC THEOREM OF CLASSICAL CALCULUS

For many years we have been guided by the idea that the "kernel" of the classical calculus is a well-known result that we call the Basic Theorem of Classical Calculus. Let us begin with its discrete analogue, which is a proposition that concerns discrete functions and appropriately conveys the spirit of the theorem.

Discrete Analogue of the Basic Theorem of Classical Calculus

Let h be a discrete function whose arguments $a_1, \ldots, a_n$ constitute an arithmetic partition of [r, s]. Then

$$\frac{G_{a_1}^{a_2} h + \cdots + G_{a_{n-1}}^{a_n} h}{n - 1} = G_r^s h \; ;$$

that is, the arithmetic average of the n - 1 successive classical gradients of h is equal to the classical gradient of h on [r, s].

The preceding result suggests the following important theorem.

Basic Theorem of Classical Calculus

If Dh is continuous on [r, s], then its arithmetic average on [r, s] is equal to the classical gradient of h on [r, s]; that is,

$$M_r^s(Dh) = G_r^s h = \frac{h(s) - h(r)}{s - r} \; .$$

In view of the preceding theorem it is appropriate to say that the arithmetic average fits naturally into the scheme of classical calculus.

1.7 THE BASIC PROBLEM OF CLASSICAL CALCULUS

Suppose that the value of a function h is known at an argument r, and suppose that f, the classical derivative of h, is continuous and known at each number in [r, s]. Find h(s).

Solution

By the Basic Theorem of Classical Calculus,

$$M_r^s f = \frac{h(s) - h(r)}{s - r} .$$

Solving for h(s), we get

$$h(s) = h(r) + (s - r) \cdot M_r^s f .$$

The number $(s - r) \cdot M_r^s f$ that appears in the preceding solution will arise with sufficient frequency to warrant a special name, "the classical integral of f on [r, s]," which is introduced in the next section.

Thus, the Basic Theorem of Classical Calculus, which involves the arithmetic average, classical derivative, and classical gradient, provides for the Basic Problem of Classical Calculus an immediate solution, which, in turn, motivates our definition of the classical integral in terms of the arithmetic average.

1.8 CLASSICAL INTEGRAL

Let f be any continuous function on [r, s]. Then the <u>classical integral of f on [r, s]</u> is denoted by $\int_r^s f$ and is defined to be the number

$$(s - r) \cdot M_r^s f .$$

We set $\int_r^r f = 0$ and $\int_s^r f = -\int_r^s f$.

The operator $\int$ is

<u>Additive</u>: $\int_r^s (f + g) = \int_r^s f + \int_r^s g$,

<u>Homogeneous</u>: $\int_r^s (c \cdot f) = c \cdot \int_r^s f$, c any constant.

The following result provides a way of expressing $\int_r^s f$ as the limit of a sequence of sums.

If f is continuous on [r, s], then $\int_r^s f$ equals the limit of the convergent sequence whose nth term is the sum

$$j_n \cdot f(a_1) + \cdots + j_n \cdot f(a_{n-1}) ,$$

where $a_1, \ldots, a_n$ is the n-fold arithmetic partition of [r, s], and j_n is the common value of

$$a_2 - a_1,\ a_3 - a_2,\ \ldots,\ a_n - a_{n-1}\ .$$

The operator $\int$ is characterized by the following three properties:

1. For any interval [r, s] and any constant function $k(x) = c$ on [r, s],

$$\int_r^s k = c \cdot (s - r)\ .$$

2. For any interval [r, s] and any functions f and g continuous on [r, s], if $f(x) \leqq g(x)$ on [r, s], then

$$\int_r^s f \leqq \int_r^s g\ .$$

3. For any numbers r, s, t such that $r < s < t$ and any function f continuous on [r, t],

$$\int_r^s f + \int_s^t f = \int_r^t f\ .$$

1.9 THE FUNDAMENTAL THEOREMS OF CLASSICAL CALCULUS

The classical derivative and classical integral are 'inversely' related in the sense indicated by the following two theorems, the second of which is a simple consequence of the Basic Theorem of Classical Calculus discussed in Section 1.6.

First Fundamental Theorem of Classical Calculus

If f is continuous on [r, s], and

$$g(x) = \int_r^x f \text{, for every number } x \text{ in } [r, s],$$

then

$$Dg = f, \quad \text{on } [r, s].$$

Second Fundamental Theorem of Classical Calculus

If Dh is continuous on [r, s], then

$$\int_r^s (Dh) = h(s) - h(r).$$

CHAPTER **2**

Weighted Classical Calculus

2.0 INTRODUCTION

A classical weight function is any function that is continuous and positive on R. There are, of course, infinitely-many classical weight functions.

For each classical weight function ω, we shall develop the ω-calculus, which is a weighted classical calculus in the sense that its operators are the classical calculus operators "weighted" by ω. It turns out that if $\omega(x) = 1$ on R, then the ω-calculus is identical with the classical calculus.

2.1 ω-MEASURE

For the remainder of this chapter, ω is an arbitrarily selected classical weight function, and $W(x) = \int_0^x \omega$ for each number x.

Notice that W is a one-to-one function on R and that $DW = \omega$.

The ω-measure of an interval [r, s] is denoted by m[r, s] and is defined to be the positive number

$$\int_r^s \omega ,$$

which equals $W(s) - W(r)$.

Notice that if $\omega(x) = 1$ on R, then the ω-measure of each interval [r, s] is equal to its classical extent s - r.

A *ω-unit interval* is any interval whose ω-measure is 1. There may or may not exist ω-unit intervals, depending on ω.

2.2 ω-AVERAGE

The definition of the ω-average of a continuous function on an interval is based on the weighted arithmetic average of n numbers, which is defined next.

Let $y_1, \ldots, y_n$ be any n numbers. Then for any n positive numbers $v_1, \ldots, v_n$, the *weighted arithmetic average of $y_1, \ldots, y_n$* is

$$\frac{v_1 y_1 + \cdots + v_n y_n}{v_1 + \cdots + v_n}.$$

Let f be any continuous function on [r, s]. Then the *ω-average of f on [r, s]* is denoted by $\underline{\underline{M}}_r^s f$ and is defined to be the limit of the convergent sequence whose nth term is the weighted arithmetic average

$$\frac{\omega(a_1)\cdot f(a_1) + \cdots + \omega(a_n)\cdot f(a_n)}{\omega(a_1) + \cdots + \omega(a_n)},$$

where $a_1, \ldots, a_n$ is the n-fold arithmetic partition of [r, s].

It turns out, of course, that $\underline{\underline{M}}_r^s f$ is the well-known weighted arithmetic average of f on [r, s]:

$$\frac{\int_r^s (w \cdot f)}{\int_r^s w} .$$

Also, $\underline{\underline{M}}_r^s f$ equals

$$\frac{\int_r^s (f \cdot DW)}{\int_r^s DW} ,$$

which is the arithmetic average of f RELATIVE to W on [r, s].

If w is constant on R, then $\underline{\underline{M}}_r^s f$ reduces to the (unweighted) arithmetic average $M_r^s f$, which equals

$$\frac{\int_r^s f}{s - r} .$$

The operator $\underline{\underline{M}}$ is

<u>Additive</u>: $\underline{\underline{M}}_r^s (f + g) = \underline{\underline{M}}_r^s f + \underline{\underline{M}}_r^s g$,

<u>Homogeneous</u>: $\underline{\underline{M}}_r^s (c \cdot f) = c \cdot \underline{\underline{M}}_r^s f$, c any constant.

The operator $\underline{\underline{M}}$ is characterized by the following three properties:

1. For any interval [r, s] and any constant function $k(x) = c$ on [r, s],

$$\underline{\underline{M}}_r^s k = c .$$

2. For any interval [r, s] and any functions f and g continuous on [r, s], if $f(x) \leqq g(x)$ on [r, s], then

$$\underline{\underline{M}}_r^s f \leqq \underline{\underline{M}}_r^s g .$$

3. For any numbers r, s, t such that $r < s < t$ and any function f continuous on [r, t],

$$m[r, s] \cdot \underline{\underline{M}}_r^s f + m[s, t] \cdot \underline{\underline{M}}_s^t f = m[r, t] \cdot \underline{\underline{M}}_r^t f .$$

A <u>ω-partition of an interval [r, s]</u> is any finite sequence of numbers $a_1, \ldots, a_n$ such that

$$r = a_1 < a_2 < \cdots < a_n = s$$

and

$$m[a_1, a_2] = m[a_2, a_3] = \cdots = m[a_{n-1}, a_n] .$$

A ω-partition with exactly n terms is said to be <u>n-fold</u>.

For each continuous function f on [r, s], we defined $\underline{\underline{M}}_r^s f$ to be the limit of the convergent sequence whose nth term is the WEIGHTED arithmetic average

$$\frac{\omega(a_1)\cdot f(a_1) + \cdots + \omega(a_n)\cdot f(a_n)}{\omega(a_1) + \cdots + \omega(a_n)},$$

where $a_1, \ldots, a_n$ is the n-fold arithmetic partition of [r, s]. Hence we were surprised to discover that $\underline{\underline{M}}_r^s f$ also equals the limit of the convergent sequence whose nth term is the (UNWEIGHTED) arithmetic average

$$\frac{f(a_1) + \cdots + f(a_n)}{n},$$

where $a_1, \ldots, a_n$ is the n-fold ω-partition of [r, s].

We close this section with a simple result worth noting.

Let f be any function continuous on [r, s], and let c be any positive constant. Then $c\omega$ is a classical weight function and on [r, s], the $c\omega$-average of f is equal to the ω-average of f.

2.3 ω-UNIFORM FUNCTIONS

In classical calculus, the linear functions are the standard to which other functions are compared; in ω-calculus the standard is provided by the ω-uniform functions, which will be defined shortly.

Recall that we defined $W(x) = \int_0^x \omega$, for each number x.

Now consider any function u such that

$$u(x) = b \cdot W(x) + c\,, \quad \text{on } R,$$

where b and c are constants.

It turns out that for any intervals $[r_1, s_1]$ and $[r_2, s_2]$ with the same ω-measure,

$$u(s_1) - u(r_1) = u(s_2) - u(r_2)\,;$$

that is, u has the same classical change on all intervals with the same ω-measure. Therefore each function on R expressible in the form $b \cdot W(x) + c$, and only such a function, is called a <u>ω-uniform function</u>.

Notice, in particular, that each ω-uniform function has the same classical change on all ω-unit intervals, if any.

Clearly every function that is constant on R is ω-uniform (take b = 0).

Moreover, if ω is constant, then the ω-uniform functions are identical with the linear functions.

In Section 2.5, we shall exploit the fact that there is exactly one ω-uniform function containing any two given points with distinct first coordinates.

It can easily be proved that if u is a ω-uniform function, then the ratio

$$\frac{u(s) - u(r)}{m[r, s]}$$

has the same value on EVERY interval [r, s]. That fact suggests the definition of ω-slope given next.

2.4 ω-SLOPE

The ω-slope of a ω-uniform function u is the common value of all the ratios

$$\frac{u(s) - u(r)}{m[r, s]} .$$

It is worth noting that the ω-slope of u is equal to its classical change on any ω-unit interval, if any.

The ω-slope of the ω-uniform function $b \cdot W(x) + c$ turns out to be b.

2.5 ω-GRADIENT

The differential branch of ω-calculus is rooted in the concept of the ω-gradient, which is defined next.

Let f be any function defined at r and s, where r < s. Then the ω-gradient of f on [r, s] is denoted by $\underline{\underline{G}}_r^s f$ and is defined to be the ω-slope of the ω-uniform function containing the points (r, f(r)) and (s, f(s)). It turns out that

$$\underline{\underline{G}}_r^s f = \frac{f(s) - f(r)}{\int_r^s \omega} = \frac{f(s) - f(r)}{m[r, s]} .$$

If u is a ω-uniform function, then $\underline{\underline{G}}_r^s u$ equals the ω-slope of u.

Since

$$\int_r^s \omega = W(s) - W(r),$$

we have

$$\underline{\underline{G}}_r^s f = \frac{f(s) - f(r)}{W(s) - W(r)},$$

which is the classical gradient of f RELATIVE to W on [r, s].

If $\omega(x) = 1$ on R, then

$$\int_r^s \omega = s - r,$$

and so

$$\underline{\underline{G}}_r^s f = G_r^s f .$$

The operator $\underline{\underline{G}}$ is

<u>Additive</u>: $\underline{\underline{G}}_r^s (f + g) = \underline{\underline{G}}_r^s f + \underline{\underline{G}}_r^s g$,

<u>Homogeneous</u>: $\underline{\underline{G}}_r^s (c \cdot f) = c \cdot \underline{\underline{G}}_r^s f$, c any constant.

Finally, when r = s, the expression for the ω-gradient yields the indeterminate form 0/0, which brings us to the next topic.

2.6 ω-DERIVATIVE

Let f be any function defined at least on an interval containing the number a in its interior. If the following limit exists, we denote it by $[\underline{\underline{D}}f](a)$ and call it the <u>ω-derivative of f at a</u>:

$$\lim_{x \to a} \frac{f(x) - f(a)}{\int_a^x \omega} .$$

Notice that if a < x, then

$$\frac{f(x) - f(a)}{\int_a^x \omega} = \underline{\underline{G}}_a^x f .$$

And if x < a, then

$$\frac{f(x) - f(a)}{\int_a^x \omega} = \underline{\underline{G}}_x^a f .$$

It can be proved that $[\underline{\underline{D}}f](a)$ and $[Df](a)$ coexist; that is, if either exists then so does the other. Moreover, if they do exist, then

$$[\underline{\underline{D}}f](a) = \frac{[Df](a)}{\omega(a)} = \frac{[Df](a)}{[DW](a)} .$$

If $\omega(x) = 1$ on R, then $[\underline{\underline{D}}f](a) = [Df](a)$.

The <u>ω-derivative of f</u>, denoted by $\underline{\underline{D}}f$, is the function that assigns to each number t the number $[\underline{\underline{D}}f](t)$, if it exists.

Since $\int_t^x \omega = W(x) - W(t)$, we have

$$[\underline{\underline{D}}f](t) = \lim_{x \to t} \frac{f(x) - f(t)}{W(x) - W(t)} .$$

Hence $\underline{\underline{D}}f$ equals the classical derivative of f RELATIVE to W.

The ω-derivative of a ω-uniform function has a constant value equal to its ω-slope. Indeed, only ω-uniform functions have ω-derivatives that are constant on R. In particular, if u is a function that is constant on R, then $\underline{\underline{D}}u$ is everywhere equal to 0.

The operator $\underline{\underline{D}}$ is

<u>Additive</u>: $\underline{\underline{D}}(f + g) = \underline{\underline{D}}f + \underline{\underline{D}}g$,

<u>Homogeneous</u>: $\underline{\underline{D}}(c \cdot f) = c \cdot \underline{\underline{D}}f$, c any constant.

In the classical calculus, Dh = h if h(x) = exp x. In the ω-calculus, $\underline{\underline{D}}h = h$ if $h(x) = \exp[W(x)]$.

It is worth noting that the ω-derivative of ω is equal to the logarithmic derivative of ω; that is,

$$\underline{\underline{D}}\omega = [D\omega]/\omega = D(\ln \omega) .$$

We conclude this section with the Mean Value Theorem of ω-Calculus. (Also provable in the ω-calculus are analogues of other classical theorems such as L'Hospital's Rule and Taylor's Theorem.)

<u>Mean Value Theorem of ω-Calculus</u>

If a function f is continuous on [r, s] and $[\underline{\underline{D}}f](x)$ exists for each number x between r and s, then between r and s there is a number c such that

$$[\underline{\underline{D}}f](c) = \underline{\underline{G}}_r^s f = \frac{f(s) - f(r)}{m[r, s]} .$$

NOTE

The following restatement of the preceding theorem is an immediate consequence of Cauchy's well-known Generalized Mean Value Theorem of Classical Calculus:

> If f is continuous on [r, s] and its classical derivative exists at every number between r and s, then between r and s there is a number c such that
>
> $$\frac{[Df](c)}{[DW](c)} = \frac{f(s) - f(r)}{W(s) - W(r)} .$$

2.7 THE BASIC THEOREM OF w-CALCULUS

Just as there is a discrete analogue of the Basic Theorem of Classical Calculus, so there is a

Discrete Analogue of the Basic Theorem of w-Calculus

> Let h be a discrete function whose arguments $a_1, \ldots, a_n$ constitute an arithmetic partition of [r, s]. For each integer i from 1 to n - 1, let
>
> $$v_i = \underset{a_i}{\overset{a_{i+1}}{M}} w .$$
>
> Then
>
> $$\frac{v_1(\underset{=}{G}{}_{a_1}^{a_2} h) + \cdots + v_{n-1}(\underset{=}{G}{}_{a_{n-1}}^{a_n} h)}{v_1 + \cdots + v_{n-1}} = \underset{=}{G}{}_r^s h ;$$
>
> that is, a certain weighted arithmetic average of the n - 1 successive w-gradients of h is equal to the w-gradient of h on [r, s].

The preceding result suggests the following important theorem.

Basic Theorem of ω-Calculus

If $\underline{\underline{D}}h$ is continuous on [r, s], then its ω-average on [r, s] is equal to the ω-gradient of h on [r, s]; that is,

$$\underline{\underline{M}}_r^s(\underline{\underline{D}}h) = \underline{\underline{G}}_r^s h = \frac{h(s) - h(r)}{m[r, s]} .$$

In view of the preceding theorem it is appropriate to say that the ω-average fits naturally into the scheme of ω-calculus.

NOTE

Here is another discrete analogue of the Basic Theorem of ω-Calculus:

Let h be a discrete function whose arguments $a_1, \ldots, a_n$ constitute a ω-partition of [r, s]. Then

$$\frac{\underline{\underline{G}}_{a_1}^{a_2} h + \cdots + \underline{\underline{G}}_{a_{n-1}}^{a_n} h}{n - 1} = \underline{\underline{G}}_r^s h ;$$

that is, the (unweighted) arithmetic average of the n - 1 successive ω-gradients of h is equal to the ω-gradient of h on [r, s].

2.8 THE BASIC PROBLEM OF ω-CALCULUS

Suppose that the value of a function h is known at an argument r, and suppose that f, the ω-derivative of h, is continuous and known at each number in [r, s]. Find h(s).

Solution

By the Basic Theorem of ω-Calculus,

$$\underline{\underline{M}}_r^s f = \frac{h(s) - h(r)}{m[r, s]} .$$

Solving for h(s), we get

$$h(s) = h(r) + m[r, s] \cdot \underline{\underline{M}}_r^s f .$$

The number $m[r, s] \cdot \underline{\underline{M}}_r^s f$ that appears in the preceding solution will arise with sufficient frequency to warrant a special name, "the ω-integral of f on [r, s]," which is introduced in the next section.

Thus, the Basic Theorem of ω-Calculus, which involves the ω-average, ω-derivative, and ω-gradient, provides for the Basic Problem of ω-Calculus an immediate solution, which, in turn, motivates our definition of the ω-integral in terms of the ω-average.

2.9 ω-INTEGRAL

Let f be any continuous function on [r, s]. Then the <u>ω-integral of f on [r, s]</u> is denoted by $\underline{\underline{\int}}_r^s f$ and is defined to be the number

$$m[r, s] \cdot \underline{\underline{M}}_r^s f \, .$$

We set $\underline{\underline{\int}}_r^r f = 0$.

It turns out that

$$\underline{\underline{\int}}_r^s f = \int_r^s (f \cdot DW) = \int_r^s (\omega \cdot f) \quad ,$$

which is a weighted classical integral, and which in certain contexts is called an inner product.

If $\omega(x) = 1$ on R, then

$$\underline{\underline{\int}}_r^s f = \int_r^s f \, .$$

The operator $\underline{\underline{\int}}$ is

<u>Additive</u>: $\underline{\underline{\int}}_r^s (f + g) = \underline{\underline{\int}}_r^s f + \underline{\underline{\int}}_r^s g$,

<u>Homogeneous</u>: $\underline{\underline{\int}}_r^s (c \cdot f) = c \cdot \underline{\underline{\int}}_r^s f$,

c any constant.

The following results provide two ways of expressing $\underline{\underline{\int}}_r^s f$ as the limit of a sequence of sums.

1. If f is continuous on [r, s], then $\underline{\underline{\int}}_r^s f$ equals the limit of the convergent sequence whose nth term is the sum

$$j_n \cdot w(a_1) \cdot f(a_1) + \cdots + j_n \cdot w(a_{n-1}) \cdot f(a_{n-1}) \ ,$$

where $a_1, \ldots, a_n$ is the n-fold arithmetic partition of [r, s], and j_n is the common value of

$$a_2 - a_1,\ a_3 - a_2,\ \ldots,\ a_n - a_{n-1} \ .$$

2. If f is continuous on [r, s], then $\underline{\underline{\int}}_r^s f$ equals the limit of the convergent sequence whose nth term is the sum

$$k_n \cdot f(a_1) + \cdots + k_n \cdot f(a_{n-1}) \ ,$$

where $a_1, \ldots, a_n$ is the n-fold w-partition of [r, s], and k_n is the common value of

$$m[a_2, a_1],\ m[a_3, a_2],\ \ldots,\ m[a_n, a_{n-1}] \ .$$

Since $k_n = W(a_{i+1}) - W(a_i)$ for $i = 1, \ldots, n-1$, it is not surprising that $\underline{\underline{\int}}_r^s f$ is the Stieltjes integral of f RELATIVE to W on [r, s].

The following two results are also worth noting.

> If f is continuous on [r, s] and c is any positive constant, then on [r, s], the $c\omega$-integral of f is equal to the product of c and the ω-integral of f.
>
> If f is continuous on [r, s] and w_1 and w_2 are classical weight functions with sum ω on R, then on [r, s], the sum of the w_1 and w_2-integrals of f is equal to the ω-integral of f.

The operator $\underline{\underline{\int}}$ is characterized by the following three properties:

1. For any interval [r, s] and any constant function $k(x) = c$ on [r, s],

$$\underline{\underline{\int}}_r^s k = c \cdot m[r, s] .$$

2. For any interval [r, s] and any functions f and g continuous on [r, s], if $f(x) \leqq g(x)$ on [r, s], then

$$\underline{\underline{\int}}_r^s f \leqq \underline{\underline{\int}}_r^s g .$$

3. For any numbers r, s, t such that $r < s < t$ and any function f continuous on [r, t],

$$\underline{\underline{\int}}_r^s f + \underline{\underline{\int}}_s^t f = \underline{\underline{\int}}_r^t f .$$

We conclude this section with another mean value theorem of ω-calculus:

If f is continuous on [r, s], then between r and s there is a number c such that

$$\underline{\underline{M}}_r^s f = f(c) \quad ;$$

that is, the ω-average of a function continuous on [r, s] is assumed at some argument between r and s.

N O T E

The following restatement of the preceding mean value theorem is an immediate consequence of the well-known Second Mean Value Theorem for Classical Integrals:

If f is continuous on [r, s], then between r and s there is a number c such that

$$\int_r^s (f \cdot \omega) = f(c) \cdot \int_r^s \omega \ .$$

2.10 THE FUNDAMENTAL THEOREMS OF ω-CALCULUS

The ω-derivative and ω-integral are 'inversely' related in the sense indicated by the following two theorems, the second of which is a simple consequence of the Basic Theorem of ω-Calculus discussed in Section 2.7.

First Fundamental Theorem of ω-Calculus

If f is continuous on [r, s], and

$$g(x) = \underline{\underline{\int}}_r^x f\,, \text{ for every number } x \text{ in } [r, s],$$

then

$$\underline{\underline{D}}g = f\,, \text{ on } [r, s].$$

Second Fundamental Theorem of ω-Calculus

If $\underline{\underline{D}}h$ is continuous on [r, s], then

$$\underline{\underline{\int}}_r^s (\underline{\underline{D}}h) = h(s) - h(r).$$

Just as the Second Fundamental Theorem of Classical Calculus is useful for evaluating classical integrals, the Second Fundamental Theorem of ω-Calculus is useful for evaluating ω-integrals. As an example, let $\omega(x) = x^2 + 1$, $f(x) = 4x$, and $h(x) = x^4 + 2x^2 + 2$, for every number x. Then $f = \underline{\underline{D}}h$, and so

$$\underline{\underline{\int}}_0^1 f = \underline{\underline{\int}}_0^1 (\underline{\underline{D}}h) = h(1) - h(0) = 5 - 2 = 3\,.$$

2.11 AN UNEXPECTED RESULT

In this section we assume that w is not identically 1 on R. (If w is identically 1 on R, then the w-calculus is the classical calculus, which does not concern us in this section.) We also assume that the reader is familiar with Chapters 5 and 6 of *Non-Newtonian Calculus*.

Recall that we defined $W(x) = \int_0^x w$, for each number x. Now assume that the range of W is R, and let $\alpha = W^{-1}$, which is a one-to-one function whose domain and range equal R. Then α generates α-arithmetic, whose realm is R.

In *Non-Newtonian Calculus* it is shown that the ordered pair of arithmetics (α-arithmetic, classical arithmetic) determines an (UNWEIGHTED) non-Newtonian calculus. Surprisingly, that system is identical with the w-calculus, which is a WEIGHTED classical calculus.

****

Since weight functions are usually assumed to be positive, we developed the w-calculus only for the case where w is a continuous <u>positive</u> function on R. However, it is possible to develop the w-calculus for the case where w is a continuous <u>negative</u> function on R.

CHAPTER 3

Weighted Non-Newtonian Calculus

3.0 INTRODUCTION

In this chapter, we shall use the notation, terminology, and results set forth in *Non-Newtonian Calculus*.

For the remainder of the chapter, α and β are arbitrarily selected generators and $*$ is the ordered pair of arithmetics (α-arithmetic, β-arithmetic). The following notations will be used.

	α-arithmetic	β-arithmetic
Realm	A	B
Addition	$\dot{+}$	$\ddot{+}$
Subtraction	$\dot{-}$	$\ddot{-}$
Multiplication	$\dot{\times}$	$\ddot{\times}$
Division	$\dot{/}$	$\ddot{/}$ or $\cdot\cdot\!-\!\!-\!\cdot\cdot$
Order	$\dot{<}$	$\ddot{<}$

We denote the isomorphism from α-arithmetic to β-arithmetic by ι.

A <u>$*$-weight function</u> is any function that is $*$-continuous and β-positive on A. (Of course, a "β-positive function" is a function whose values are all β-positive numbers.) There are infinitely-many $*$-weight functions.

For each $\star$-weight function ω, we shall develop the $\omega\star$-calculus, which is a weighted $\star$-calculus in the sense that its operators are the $\star$-calculus operators "weighted" by ω. It turns out that if $\omega(x) = \ddot{1}$ on A, then the $\omega\star$-calculus is identical with the $\star$-calculus. Also, in the special case for which $\alpha(x) = x = \beta(x)$ on R, the $\omega\star$-calculus turns out to be a weighted classical calculus.

The operators of each $\omega\star$-calculus are applied only to functions with arguments in A and values in B. Accordingly, unless indicated or implied otherwise, all functions considered in this chapter are assumed to be of that character.

We close this section with a notational convention, which will be used shortly.

For each $\star$-continuous function f on an α-interval $\dot{[}r, s\dot{]}$,

$$\overset{\star}{\int_s^r} f \quad \text{denotes} \quad \ddot{0} \ddot{-} \overset{\star}{\int_r^s} f \; .$$

3.1 $\omega\star$-MEASURE

For the remainder of this chapter, ω is an arbitrarily selected $\star$-weight function, and

$$W(x) = \overset{\star}{\int_{\dot{0}}^x} \omega \, ,$$

for each number x in A.

Then W is a one-to-one function on A, and $\overset{\star}{D}W = \omega$.

The <u>ω*-measure</u> of an α-interval $[r, s]$ is denoted by $\overset{*}{m}[r, s]$ and is defined to be the β-positive number

$$\overset{*}{\int_r^s} \omega ,$$

which equals $W(s) \ddot{-} W(r)$.

A <u>ω*-unit interval</u> is any α-interval whose ω*-measure is $\ddot{1}$. There may or may not exist ω*-unit intervals, depending on ω.

3.2 ω*-AVERAGE

Our definition of the ω*-average of a *-continuous function on an α-interval is based on the weighted β-average of n numbers, which we define next.

Let $y_1, \ldots, y_n$ be any n numbers in B. Then for any n β-positive numbers $v_1, \ldots, v_n$, the <u>weighted β-average of $y_1, \ldots, y_n$</u> is the following number in B:

$$\ddot{\ }\frac{(v_1 \ddot{\times} y_1) \ddot{+} \cdots \ddot{+} (v_n \ddot{\times} y_n)}{v_1 \ddot{+} \cdots \ddot{+} v_n}\ddot{\ } .$$

Let f be any function *-continuous on $[r, s]$. Then the <u>ω*-average of f on $[r, s]$</u> is denoted by $\overset{*}{\underline{\underline{M}}}{}_r^s f$ and is defined to be the β-limit of the β-convergent sequence whose nth term is the weighted β-average

$$\ddot{\ }\frac{[\omega(a_1) \ddot{\times} f(a_1)] \ddot{+} \cdots \ddot{+} [\omega(a_n) \ddot{\times} f(a_n)]}{\omega(a_1) \ddot{+} \cdots \ddot{+} \omega(a_n)}\ddot{\ } ,$$

where $a_1, \ldots, a_n$ is the n-fold α-partition of $[r, s]$.

It turns out that $\overset{*}{\underline{\underline{M}}}{}_r^s f$ equals

$$\cdot\cdot\frac{\overset{*}{\int}_r^s (\omega \ddot{\times} f)}{\overset{*}{\int}_r^s \omega}\cdot\cdot \quad .$$

Since

$$\overset{*}{\underline{\underline{M}}}{}_r^s f = \cdot\cdot\frac{\overset{*}{\int}_r^s (f \ddot{\times} \overset{*}{D}W)}{\overset{*}{\int}_r^s \overset{*}{D}W}\cdot\cdot \quad ,$$

we may call $\overset{*}{\underline{\underline{M}}}{}_r^s f$ "the ∗-average of f RELATIVE to W on $[r, s]$."

If ω is constant on A, then $\overset{*}{\underline{\underline{M}}}{}_r^s f$ reduces to the (unweighted) ∗-average $\overset{*}{M}{}_r^s f$.

In certain special cases the ω∗-average is well-known. For example, if $\alpha(x) = x = \beta(x)$ on R, then $\overset{*}{\underline{\underline{M}}}{}_r^s f$ is the weighted arithmetic average of f on [r, s] (see Section 2.2). If $\alpha(x) = x$ and $\beta(x) = \exp x$, on R, then $\overset{*}{\underline{\underline{M}}}{}_r^s f$ is the weighted geometric average of f on [r, s]. Also, if $\alpha(x) = x$ on R,

$$\beta(x) = \begin{cases} x^p & \text{for } x > 0 \\ 0 & \text{for } x = 0 \\ -(-x)^p & \text{for } x < 0 \end{cases} , \qquad p \neq 0 ,$$

and f is positive, then $\overset{*}{\underline{\underline{M}}}{}_r^s f$ is the weighted pth-power average of f on [r, s], which, if $p = -1$, is the weighted harmonic average of f on [r, s].

The operator $\underline{\underline{\overset{*}{M}}}$ is

<u>β-Additive</u>: $\qquad \underline{\underline{\overset{*}{M}}}{}_r^s(f \ddot{+} g) = \underline{\underline{\overset{*}{M}}}{}_r^s f \ddot{+} \underline{\underline{\overset{*}{M}}}{}_r^s g$,

<u>β-Homogeneous</u>: $\qquad \underline{\underline{\overset{*}{M}}}{}_r^s(c \ddot{\times} f) = c \ddot{\times} \underline{\underline{\overset{*}{M}}}{}_r^s f$,

c any constant in B.

The operator $\underline{\underline{\overset{*}{M}}}$ is characterized by the following three properties:

1. For any α-interval $\dot{[}r, s\dot{]}$ and any constant function $k(x) = c$ on $\dot{[}r, s\dot{]}$,

$$\underline{\underline{\overset{*}{M}}}{}_r^s k = c \ .$$

2. For any α-interval $\dot{[}r, s\dot{]}$ and any functions f and g $*$-continuous on $\dot{[}r, s\dot{]}$, if $f(x) \ddot{\leqq} g(x)$ on $\dot{[}r, s\dot{]}$, then

$$\underline{\underline{\overset{*}{M}}}{}_r^s f \ddot{\leqq} \underline{\underline{\overset{*}{M}}}{}_r^s g \ .$$

3. For any numbers r, s, t in A such that $r \dot{<} s \dot{<} t$ and any function f $*$-continuous on $\dot{[}r, t\dot{]}$,

$$(\overset{*}{m}\dot{[}r, s\dot{]} \ddot{\times} \underline{\underline{\overset{*}{M}}}{}_r^s f) \ddot{+} (\overset{*}{m}\dot{[}s, t\dot{]} \ddot{\times} \underline{\underline{\overset{*}{M}}}{}_s^t f) = \overset{*}{m}\dot{[}r, t\dot{]} \ddot{\times} \underline{\underline{\overset{*}{M}}}{}_r^t f \ .$$

A <u>*ω*$*$-partition of an α-interval</u> $\dot{[}r, s\dot{]}$ is any finite sequence of numbers $a_1, \ldots, a_n$ in A such that

$$r = a_1 \dot{<} a_2 \dot{<} \cdots \dot{<} a_n = s$$

and

$$\overset{*}{m}\dot{[}a_1, a_2\dot{]} = \overset{*}{m}\dot{[}a_2, a_3\dot{]} = \cdots = \overset{*}{m}\dot{[}a_{n-1}, a_n\dot{]} \ .$$

A *ω*$*$-partition with exactly n terms is said to be <u>n-fold</u>.

For each function f $*$-continuous on [r, s], we defined $\overset{*}{\underline{\underline{M}}}{}_r^s f$ to be the β-limit of the β-convergent sequence whose nth term is the WEIGHTED β-average

$$\cdot\cdot\frac{[w(a_1)\ \ddot{\times}\ f(a_1)]\ \ddot{+}\ \cdots\ \ddot{+}\ [w(a_n)\ \ddot{\times}\ f(a_n)]}{w(a_1)\ \ddot{+}\ \cdots\ \ddot{+}\ w(a_n)}\cdot\cdot\ ,$$

where $a_1, \ldots, a_n$ is the n-fold α-partition of [r, s]. Nevertheless, it turns out that $\overset{*}{\underline{\underline{M}}}{}_r^s f$ also equals the β-limit of the β-convergent sequence whose nth term is the (UNWEIGHTED) β-average

$$\cdot\cdot\frac{f(a_1)\ \ddot{+}\ \cdots\ \ddot{+}\ f(a_n)}{\beta(n)}\cdot\cdot\ ,$$

where $a_1, \ldots, a_n$ is the n-fold $w*$-partition of [r, s].

We close this section with a simple result worth noting.

Let f be any function $*$-continuous on [r, s], let c be any β-positive constant, and let $v = c\ \ddot{\times}\ w$. Then v is a $*$-weight function and on [r, s], the $v*$-average of f is equal to the $w*$-average of f.

3.3 $w*$-UNIFORM FUNCTIONS

In the $w*$-calculus, the $w*$-uniform functions, which will be defined shortly, are the standard to which other functions are compared.

Recall that we defined $W(x) = \overset{*}{\int}_{\dot{0}}^{x} w$ for each number x in A.

Now consider any function u such that

$$u(x) = [b \ddot{\times} \mathcal{W}(x)] \ddot{+} c\,, \quad \text{on } A\,,$$

where b and c are constants in B.

It turns out that for any α-intervals $\dot{[}r_1, s_1\dot{]}$ and $\dot{[}r_2, s_2\dot{]}$ with the same w*-measure,

$$u(s_1) \ddot{-} u(r_1) = u(s_2) \ddot{-} u(r_2)\,;$$

that is, u has the same β-change on all α-intervals with the same w*-measure. Therefore each function on A expressible in the form $[b \ddot{\times} \mathcal{W}(x)] \ddot{+} c$, and only such a function, is called a <u>w*-uniform function</u>.

Notice, in particular, that each w*-uniform function has the same β-change on all w*-unit intervals, if any.

Clearly every function that is constant on A is w*-uniform (take $b = \ddot{0}$).

Moreover, if w is constant, then the w*-uniform functions are identical with the *-uniform functions.

In Section 3.5, we shall exploit the fact that there is exactly one w*-uniform function containing any two given *-points with distinct first coordinates.

It can easily be proved that if u is a w*-uniform function, then the β-ratio

$$\frac{u(s) \ddot{-} u(r)}{\overset{*}{m}\dot{[}r, s\dot{]}}$$

has the same value on EVERY α-interval $\dot{[}r, s\dot{]}$. That fact suggests the definition of w*-slope given next.

3.4 ω*-SLOPE

The ω*-slope of a ω*-uniform function u is the common value of all the β-ratios

$$\ddot{\cdot\cdot}\,\frac{u(s) \ddot{-} u(r)}{\overset{*}{m}[\dot{r}, \dot{s}]}\,\ddot{\cdot\cdot} \quad .$$

It is worth noting that the ω*-slope of u is equal to its β-change on any ω*-unit interval, if any.

The ω*-slope of the ω*-uniform function $[b \ddot{\times} W(x)] \ddot{+} c$ turns out to be b.

3.5 ω*-GRADIENT

The differential branch of ω*-calculus is rooted in the concept of ω*-gradient, which is defined next.

Let f be any function defined at r and s, where $r \dot{<} s$. Then the ω*-gradient of f on $[\dot{r}, \dot{s}]$ is denoted by $\overset{*}{\underline{\underline{G}}}{}_r^s f$ and is defined to be the ω*-slope of the ω*-uniform function containing the *-points (r, f(r)) and (s, f(s)). It turns out that

$$\overset{*}{\underline{\underline{G}}}{}_r^s f = \ddot{\cdot\cdot}\,\frac{f(s) \ddot{-} f(r)}{\overset{*}{\int_r^s} \omega}\,\ddot{\cdot\cdot} = \ddot{\cdot\cdot}\,\frac{f(s) \ddot{-} f(r)}{\overset{*}{m}[\dot{r}, \dot{s}]}\,\ddot{\cdot\cdot} \quad .$$

If u is a ω*-uniform function, then $\overset{*}{\underline{\underline{G}}}{}_r^s u$ equals the ω*-slope of u.

Since $\displaystyle \overset{*}{\int_r^s} w = W(s) \ddot{-} W(r)$,

we have

$$\overset{*}{\underline{\underline{G}}}{}_r^s f = \ddot{\frac{f(s) \ddot{-} f(r)}{W(s) \ddot{-} W(r)}} ,$$

and so we may call $\overset{*}{\underline{\underline{G}}}{}_r^s f$ "the $*$-gradient of f RELATIVE to W on $[\dot{r}, \dot{s}]$."

If $w(x) = \ddot{1}$ on A, then

$$\overset{*}{\int_r^s} w = \iota(s) \ddot{-} \iota(r) ,$$

and so $\overset{*}{\underline{\underline{G}}}{}_r^s f$ equals the $*$-gradient of f on $[\dot{r}, \dot{s}]$.

The operator $\overset{*}{\underline{\underline{G}}}$ is

<u>β-Additive</u>: $\overset{*}{\underline{\underline{G}}}{}_r^s (f \ddot{+} g) = \overset{*}{\underline{\underline{G}}}{}_r^s f \ddot{+} \overset{*}{\underline{\underline{G}}}{}_r^s g$,

<u>β-Homogeneous</u>: $\overset{*}{\underline{\underline{G}}}{}_r^s (c \ddot{\times} f) = c \ddot{\times} \overset{*}{\underline{\underline{G}}}{}_r^s f$,

c any constant in B.

Finally, when r = s, the expression for the $w*$-gradient yields the indeterminate form $\ddot{0} \ddot{/} \ddot{0}$, which brings us to the next topic.

3.6 w*-DERIVATIVE

Let f be any function defined at least on an α-interval containing the number a in its α-interior. If the following *-limit exists, we denote it by $[\overset{*}{\underline{\underline{D}}}f](a)$ and call it the <u>w*-derivative of f at a</u>:

$$\underset{x\to a}{*\text{-}\lim} \;\cdot\cdot\frac{f(x) \ddot{-} f(a)}{\overset{*}{\int_a^x} w}\cdot\cdot \quad .$$

Notice that if a $\dot{<}$ x, then

$$\cdot\cdot\frac{f(x) \ddot{-} f(a)}{\overset{*}{\int_a^x} w}\cdot\cdot = \overset{*x}{\underset{a}{\underline{\underline{G}}}} f \;.$$

And if x $\dot{<}$ a, then

$$\cdot\cdot\frac{f(x) \ddot{-} f(a)}{\overset{*}{\int_a^x} w}\cdot\cdot = \overset{*a}{\underset{x}{\underline{\underline{G}}}} f \;.$$

If it exists, the number $[\overset{*}{\underline{\underline{D}}}f](a)$ is in B.

It can be proved that $[\overset{*}{\underline{\underline{D}}}f](a)$ and $[\overset{*}{D}f](a)$ coexist. Moreover, if they do exist, then

$$[\overset{*}{\underline{\underline{D}}}f](a) = \cdot\cdot\frac{[\overset{*}{D}f](a)}{w(a)}\cdot\cdot = \cdot\cdot\frac{[\overset{*}{D}f](a)}{[\overset{*}{D}W](a)}\cdot\cdot \quad .$$

If $w(x) = \ddot{1}$ on A, then $[\overset{*}{\underline{\underline{D}}}f](a) = [\overset{*}{D}f](a)$.

The $\omega\star$-derivative of f, denoted by $\overset{\star}{\underline{\underline{D}}}f$, is the function that assigns to each number t in A the number $[\overset{\star}{\underline{\underline{D}}}f](t)$, if it exists.

Since $\overset{\star}{\int}_t^x \omega = W(x) \ddot{-} W(t)$, we have

$$[\overset{\star}{\underline{\underline{D}}}f](t) = \star\text{-}\lim_{x\to t} \cdot\cdot \frac{f(x) \ddot{-} f(t)}{W(x) \ddot{-} W(t)} \cdot\cdot \quad .$$

Hence $\overset{\star}{\underline{\underline{D}}}f$ may be called "the $\star$-derivative of f RELATIVE to W."

The $\omega\star$-derivative of a $\omega\star$-uniform function has a constant value equal to its $\omega\star$-slope. Indeed, only $\omega\star$-uniform functions have $\omega\star$-derivatives that are constant on A. In particular, if u is a function that is constant on A, then $\overset{\star}{\underline{\underline{D}}}u = \ddot{0}$ on A.

The operator $\overset{\star}{\underline{\underline{D}}}$ is

<u>β-Additive</u>: $\overset{\star}{\underline{\underline{D}}}(f \ddot{+} g) = \overset{\star}{\underline{\underline{D}}}f \ddot{+} \overset{\star}{\underline{\underline{D}}}g$,

<u>β-Homogeneous</u>: $\overset{\star}{\underline{\underline{D}}}(c \ddot{\times} f) = c \ddot{\times} \overset{\star}{\underline{\underline{D}}}f$,

c any constant in B.

It is worth noting that if $\alpha(x) = \exp x$ and $\beta(x) = x$, on R, then the $\omega\star$-derivative of ω is equal to the so-called elasticity of ω. (The <u>elasticity</u> at an argument a of a function f with positive arguments and values is defined to be

$$\lim_{x\to a} \frac{[f(x) - f(a)]/f(a)}{(x - a)/a} \quad ,$$

which, if it exists, equals $a\cdot[Df](a)/f(a)$. Elasticity is often used in economics.)

We conclude this section with the Mean Value Theorem of $\omega\star$-Calculus. (Also provable in the $\omega\star$-calculus are analogues of other classical theorems such as L'Hospital's Rule and Taylor's Theorem.)

Mean Value Theorem of $\omega\star$-Calculus

If a function f is $\star$-continuous on $\dot{[}r, s\dot{]}$ and $[\overset{\star}{\underline{\underline{D}}}f](x)$ exists for each number x such that $r \dot{<} x \dot{<} s$, then there exists a number c for which $r \dot{<} c \dot{<} s$ and

$$[\overset{\star}{\underline{\underline{D}}}f](c) = \overset{\star}{\underline{\underline{G}}}{}_r^s f = \cdot\cdot\frac{f(s) \ddot{-} f(r)}{\overset{\star}{m}\dot{[}r, s\dot{]}}\cdot\cdot \quad .$$

3.7 THE BASIC THEOREM OF $\omega\star$-CALCULUS

We begin with a discrete analogue. (Another discrete analogue is provided in the Note at the end of this section.)

Discrete Analogue of the Basic Theorem of $\omega\star$-Calculus

Let h be a discrete function whose arguments $a_1, \ldots, a_n$ constitute an α-partition of $\dot{[}r, s\dot{]}$. For each integer i from 1 to n - 1, let

$$v_i = \overset{\star}{M}{}_{a_i}^{a_{i+1}} \omega \; .$$

Then

$$\cdot\cdot\frac{(v_1 \ddot{\times} \overset{\star}{\underline{\underline{G}}}{}_{a_1}^{a_2} h) \ddot{+} \cdots \ddot{+} (v_{n-1} \ddot{\times} \overset{\star}{\underline{\underline{G}}}{}_{a_{n-1}}^{a_n} h)}{v_1 \ddot{+} \cdots \ddot{+} v_{n-1}}\cdot\cdot = \overset{\star}{\underline{\underline{G}}}{}_r^s h \; ;$$

that is, a certain weighted β-average of the n - 1 successive $\omega *$-gradients of h is equal to the $\omega *$-gradient of h on [r, s].

The preceding result suggests the following important theorem.

<u>Basic Theorem of $\omega *$-Calculus</u>

If $\overset{*}{\underline{\underline{D}}}h$ is $*$-continuous on [r, s], then its $\omega *$-average on [r, s] is equal to the $\omega *$-gradient of h on [r, s]; that is,

$$\overset{*}{\underline{\underline{M}}}{}_r^s(\overset{*}{\underline{\underline{D}}}h) = \overset{*}{\underline{\underline{G}}}{}_r^s h = \cdot\cdot\frac{h(s) \ddot{-} h(r)}{\overset{*}{m}[r, s]}\cdot\cdot \quad .$$

In view of the preceding theorem it is appropriate to say that the $\omega *$-average fits naturally into the scheme of $\omega *$-calculus.

<u>N O T E</u>

Here is another discrete analogue of the Basic Theorem of $\omega *$-Calculus:

Let h be a discrete function whose arguments $a_1, \ldots, a_n$ constitute a $\omega *$-partition of [r, s]. Then

$$\cdot\cdot\frac{\overset{*}{\underline{\underline{G}}}{}_{a_1}^{a_2} h \ddot{+} \cdots \ddot{+} \overset{*}{\underline{\underline{G}}}{}_{a_{n-1}}^{a_n} h}{\beta(n - 1)}\cdot\cdot = \overset{*}{\underline{\underline{G}}}{}_r^s h \;;$$

that is, the (unweighted) β-average of the n - 1 successive $\omega *$-gradients of h is equal to the $\omega *$-gradient of h on [r, s].

3.8 THE BASIC PROBLEM OF $\omega\star$-CALCULUS

Suppose that the value of a function h is known at an argument r, and suppose that f, the $\omega\star$-derivative of h, is $\star$-continuous and known at each number in $\dot{[}r, s\dot{]}$. Find h(s).

<u>Solution</u>

By the Basic Theorem of $\omega\star$-Calculus,

$$\overset{\star}{\underline{\underline{M}}}{}_r^s f = \cdot\cdot\frac{h(s) \ddot{-} h(r)}{\overset{\star}{m}\dot{[}r, s\dot{]}}\cdot\cdot \quad .$$

Solving for h(s), we get

$$h(s) = h(r) \ddot{+} (\overset{\star}{m}\dot{[}r, s\dot{]} \ddot{\times} \overset{\star}{\underline{\underline{M}}}{}_r^s f) .$$

The number $\overset{\star}{m}\dot{[}r, s\dot{]} \ddot{\times} \overset{\star}{\underline{\underline{M}}}{}_r^s f$ that appears in the preceding solution will arise with sufficient frequency to warrant a special name, "the $\omega\star$-integral of f on $\dot{[}r, s\dot{]}$," which is introduced in the next section.

Thus, the Basic Theorem of $\omega\star$-Calculus, which involves the $\omega\star$-average, $\omega\star$-derivative, and $\omega\star$-gradient, provides for the Basic Problem of $\omega\star$-Calculus an immediate solution, which, in turn, motivates our definition of the $\omega\star$-integral in terms of the $\omega\star$-average.

3.9 w*-INTEGRAL

Let f be any function *-continuous on $[r, \dot{s}]$. Then the <u>w*-integral of f on $[r, \dot{s}]$</u> is denoted by $\underline{\underline{\int}}_r^{*s} f$ and is defined to be the following number in B:

$$\overset{*}{m}[r, \dot{s}] \;\ddot{\times}\; \underline{\underline{M}}_r^{*s} f \;.$$

We set $\underline{\underline{\int}}_r^{*r} f = \ddot{0}$.

It turns out that

$$\underline{\underline{\int}}_r^{*s} f = \int_r^{*s} (w \;\ddot{\times}\; f) \;,$$

which might appropriately be called "a weighted *-integral."

Since

$$\underline{\underline{\int}}_r^{*s} f = \int_r^{*s} (f \;\ddot{\times}\; \overset{*}{D}W) \;,$$

we may call $\underline{\underline{\int}}_r^{*s} f$ "the *-integral of f RELATIVE to W on $[r, \dot{s}]$."

If $w(x) = \ddot{1}$ on A, then

$$\underline{\underline{\int}}_r^{*s} f = \int_r^{*s} f \;.$$

The operator $\underset{=}{\int}^{\star}$ is

<u>β-Additive</u>: $\quad \underset{=}{\int}_{r}^{\star s} (f \ddot{+} g) = \underset{=}{\int}_{r}^{\star s} f \ddot{+} \underset{=}{\int}_{r}^{\star s} g ,$

<u>β-Homogeneous</u>: $\quad \underset{=}{\int}_{r}^{\star s} (c \ddot{\times} f) = c \ddot{\times} \underset{=}{\int}_{r}^{\star s} f ,$

c any constant in B.

The following results provide two ways of expressing $\underset{=}{\int}_{r}^{\star s} f$ as the β-limit of a sequence of β-sums.

1. If f is ⋆-continuous on $\dot{[}r, s\dot{]}$, then $\underset{=}{\int}_{r}^{\star s} f$ equals the β-limit of the β-convergent sequence whose nth term is the β-sum

$$[\iota(j_n) \ddot{\times} \omega(a_1) \ddot{\times} f(a_1)] \ddot{+} \cdots \ddot{+} [\iota(j_n) \ddot{\times} \omega(a_{n-1}) \ddot{\times} f(a_{n-1})] ,$$

where $a_1, \ldots, a_n$ is the n-fold α-partition of $\dot{[}r, s\dot{]}$, and j_n is the common value of

$$a_2 \dot{-} a_1, a_3 \dot{-} a_2, \ldots, a_n \dot{-} a_{n-1} .$$

2. If f is ⋆-continuous on $\dot{[}r, s\dot{]}$, then $\underset{=}{\int}_{r}^{\star s} f$ equals the β-limit of the β-convergent sequence whose nth term is the β-sum

$$[k_n \ddot{\times} f(a_1)] \ddot{+} \cdots \ddot{+} [k_n \ddot{\times} f(a_{n-1})] ,$$

where $a_1, \ldots, a_n$ is the n-fold ω⋆-partition of $\dot{[}r, s\dot{]}$, and k_n is the common value of

$$\overset{\star}{m}\dot{[}a_1, a_2\dot{]}, \overset{\star}{m}\dot{[}a_2, a_3\dot{]}, \ldots, \overset{\star}{m}\dot{[}a_{n-1}, a_n\dot{]} .$$

The following two results are also worth noting.

> If f is $\star$-continuous on $\dot{[}r, s\dot{]}$, c is any β-positive constant, and $v = c \ddot{\times} w$, then on $\dot{[}r, s\dot{]}$, the v$\star$-integral of f is equal to the β-product of c and the $w\star$-integral of f.
>
> If f is $\star$-continuous on $\dot{[}r, s\dot{]}$ and u and v are $\star$-weight functions with β-sum w on A, then on $\dot{[}r, s\dot{]}$, the β-sum of the u$\star$ and v$\star$-integrals of f is equal to the $w\star$-integral of f.

The operator $\underset{\equiv}{\overset{\star}{\int}}$ is characterized by the following three properties:

1. For any α-interval $\dot{[}r, s\dot{]}$ and any constant function k(x) = c on $\dot{[}r, s\dot{]}$,

$$\underset{\equiv}{\overset{\star}{\int}}{}_r^s k = c \ddot{\times} \overset{\star}{m}\dot{[}r, s\dot{]} \; .$$

2. For any α-interval $\dot{[}r, s\dot{]}$ and any functions f and g $\star$-continuous on $\dot{[}r, s\dot{]}$, if $f(x) \ddot{\leqq} g(x)$ on $\dot{[}r, s\dot{]}$, then

$$\underset{\equiv}{\overset{\star}{\int}}{}_r^s f \ddot{\leqq} \underset{\equiv}{\overset{\star}{\int}}{}_r^s g \; .$$

3. For any numbers r, s, t in A such that $r \dot{<} s \dot{<} t$ and any function f $\star$-continuous on $\dot{[}r, t\dot{]}$,

$$\underset{\equiv}{\overset{\star}{\int}}{}_r^s f \ddot{+} \underset{\equiv}{\overset{\star}{\int}}{}_s^t f = \underset{\equiv}{\overset{\star}{\int}}{}_r^t f \; .$$

We conclude this section with another mean value theorem of ω*-calculus:

> If f is *-continuous on $\dot{[}r, s\dot{]}$, then there is a number c in A such that $r \dot{<} c \dot{<} s$ and
>
> $$\overset{*}{\underline{\underline{M}}}{}_r^s f = f(c) ;$$
>
> that is, the ω*-average of a function *-continuous on $\dot{[}r, s\dot{]}$ is assumed at some argument.

3.10 THE FUNDAMENTAL THEOREMS OF ω*-CALCULUS

The ω*-derivative and ω*-integral are 'inversely' related in the sense indicated by the following two theorems, the second of which is a simple consequence of the Basic Theorem of ω*-Calculus discussed in Section 3.7.

<u>First Fundamental Theorem of ω*-Calculus</u>

If f is *-continuous on $\dot{[}r, s\dot{]}$, and

$$g(x) = \underline{\underline{\int}}{}_r^{*x} f , \quad \text{for every number } x \text{ in } \dot{[}r, s\dot{]} ,$$

then

$$\overset{*}{\underline{\underline{D}}}g = f , \quad \text{on } \dot{[}r, s\dot{]} .$$

<u>Second Fundamental Theorem of ω*-Calculus</u>

If $\overset{*}{\underline{\underline{D}}}h$ is *-continuous on $\dot{[}r, s\dot{]}$, then

$$\underline{\underline{\int}}{}_r^{*s} (\overset{*}{\underline{\underline{D}}}h) = h(s) \ddot{-} h(r) .$$

3.11 RELATIONSHIP BETWEEN ω*-CALCULUS AND WEIGHTED CLASSICAL CALCULUS

For the remainder of this section, let $\bar{\omega}$ be the classical weight function such that $\bar{\omega}(t) = \beta^{-1}(\omega(\alpha(t)))$ for each number t. We shall indicate the uniform relationships between the corresponding operators of ω*-calculus and $\bar{\omega}$-calculus.

For each number x in A, let $\bar{x} = \alpha^{-1}(x)$. Let f be a function with arguments in A and values in B, and set $\bar{f}(t) = \beta^{-1}(f(\alpha(t)))$.

We have the following uniform relationships:

$$(1) \qquad \overset{*}{\underline{\underline{G}}}{}_r^s f = \beta\left\{ \underline{\underline{G}}_{\bar{r}}^{\bar{s}} \bar{f} \right\},$$

$$(2) \qquad [\overset{*}{\underline{\underline{D}}} f](a) = \beta\left\{ [\underline{\underline{D}}\bar{f}](\bar{a}) \right\},$$

$$(3) \qquad \overset{*}{\underline{\underline{M}}}{}_r^s f = \beta\left\{ \underline{\underline{M}}_{\bar{r}}^{\bar{s}} \bar{f} \right\},$$

$$(4) \qquad \overset{*}{\int_{\underline{\underline{r}}}^{s}} f = \beta\left\{ \int_{\underline{\underline{\bar{r}}}}^{\bar{s}} \bar{f} \right\}.$$

(In (2), the derivatives $[\overset{*}{\underline{\underline{D}}} f](a)$ and $[\underline{\underline{D}}\bar{f}](\bar{a})$ coexist; in (3) and (4), we assume f is *-continuous on $[r, s]$.)

It is also worth noting that f is a ω*-uniform function if and only if $\bar{f}$ is a $\bar{\omega}$-uniform function.

The foregoing results suggest that for each theorem in $\bar{\omega}$-calculus, there is a corresponding theorem in ω*-calculus, and conversely.

3.12 AN ALTERNATIVE APPROACH TO $\omega\ast$-CALCULUS

First recall that we defined $W(x) = \overset{\ast}{\int_{0}^{x}} \omega$, for each number x in A. Now assume that the range of W is B, and let $\bar{\alpha} = W^{-1} \circ \beta$, which is a one-to-one function whose domain is R and range is A. Then $\bar{\alpha}$ generates $\bar{\alpha}$-arithmetic, whose realm is A.

Let $\bar{\ast}$ be the ordered pair of arithmetics ($\bar{\alpha}$-arithmetic, β-arithmetic). It can be proved that the operators of $\bar{\ast}$-calculus are identical with the corresponding operators of $\omega\ast$-calculus and that the $\bar{\ast}$-uniform functions are identical with the $\omega\ast$-uniform functions. Indeed, $\bar{\ast}$-calculus is identical with $\omega\ast$-calculus. Thus, the $\omega\ast$-calculus, which is a WEIGHTED $\ast$-calculus, can be reformulated as an UNWEIGHTED calculus.

Accordingly, in *Non-Newtonian Calculus* all the discussions, developments, results, interpretations, and heuristic principles that pertain to the $\bar{\ast}$-calculus are pertinent to the $\omega\ast$-calculus.

3.13 WEIGHTED GEOMETRIC CALCULUS

In this final section we shall discuss a certain system of weighted calculus that has many attractive features and may prove to be one of the most important of all the systems of calculus.

Henceforth, $\alpha(x) = x$ and $\beta(x) = \exp x$, on R, and $\ast$ is the ordered pair of arithmetics (α-arithmetic, β-arithmetic). Let ω be an arbitrarily selected $\ast$-weight function.

The $\omega\ast$-calculus is a weighted geometric calculus in the sense that its operators are the geometric calculus operators "weighted" by ω.

The $w*$-measure of an interval [r, s] equals

$$\exp\left\{\int_r^s \ln w\right\} .$$

The $w*$-average of a continuous positive function f on [r, s] equals

$$\exp\left\{\frac{\int_r^s (\ln w \cdot \ln f)}{\int_r^s \ln w}\right\} ,$$

which is the well-known weighted geometric average of f on [r, s].

The $w*$-uniform functions are expressible in the form

$$\exp\left\{(b \cdot \int_0^x \ln w) + c\right\} ,$$

where b and c are constants and x is unrestricted in R. The $w*$-slope of the preceding function equals exp b .

The $w*$-gradient of a positive function f on [r, s] equals

$$\left\{\frac{f(s)}{f(r)}\right\}^{1/t} ,$$

where $t = \int_r^s \ln w$.

The ω*-derivative of a positive function f equals

$$\exp\left\{\frac{D(\ln f)}{\ln \omega}\right\},$$

which is constant on R if and only if f is a ω*-uniform function.

The ω*-integral of a continuous positive function f on [r, s] equals

$$\exp\left\{\int_r^s (\ln \omega \cdot \ln f)\right\}.$$

All the operators of the ω*-calculus are multiplicative and involutional; for example,

$$\overset{*}{\underline{\underline{D}}}(f \cdot g) = \overset{*}{\underline{\underline{D}}}f \cdot \overset{*}{\underline{\underline{D}}}g \quad \text{and}$$

$$\overset{*}{\underline{\underline{D}}}(f^c) = (\overset{*}{\underline{\underline{D}}}f)^c, \quad c \text{ any constant}.$$

If $\omega(x) = e$ on R, then the operators of ω*-calculus are identical with the corresponding operators of the geometric calculus, the ω*-uniform functions are identical with the geometrically-uniform functions, and, indeed, ω*-calculus is identical with geometric calculus.

****

In this chapter we developed the ω*-calculus only for the case where ω is a *-continuous <u>β-positive</u> function on A. However, it is possible to develop the ω*-calculus for the case where ω is a *-continuous <u>β-negative</u> function on A.

LIST OF SYMBOLS

I N D E X

In the classical calculus the line is used as a standard against which other curves are compared. In the following remarks, Roger Joseph Boscovich (1711 - 1787) may have been the only person to anticipate a fundamental idea in non-Newtonian calculus: *a nonlinear curve may be used as a standard against which other curves (including lines) can be compared.*

> 'But if some mind very different from ours were to look upon some property of some curved line as we do on the evenness of a straight line, he would not recognize as such the evenness of a straight line; nor would he arrange the elements of his geometry according to that very different system, and would investigate quite other relationships as I have suggested in my notes.
>
> 'We fashion our geometry on the properties of a straight line because that seems to us to be the simplest of all. But really all lines that are continuous and of a uniform nature are just as simple as one another. Another kind of mind which might form an equally clear mental perception of some property of any one of these curves, as we do of the congruence of a straight line, might believe these curves to be the simplest of all, and from that property of these curves build up the elements of a very different geometry, referring all other curves to that one, just as we compare them to a straight line. Indeed, these minds, if they noticed and formed an extremely clear perception of some property of, say, the parabola, would not seek, as our geometers do, to *rectify* the parabola, they would endeavour, if one may coin the expression, to *parabolify* the straight line.'

The quotation above appears in Dr. J. F. Scott's article "Boscovich's Mathematics," which is one of nine articles in *Roger Joseph Boscovich,* edited by Lancelot Law Whyte and first published by George Allen & Unwin in 1961.

Grossman, M. and Katz, R. *Non-Newtonian Calculus.* Rockport, MA: Mathco, 1972.
Included in this book, which was the first publication on non-Newtonian calculus, are discussions of nine specific non-Newtonian calculi, the general theory of non-Newtonian calculus, and heuristic guides for the application thereof.

Grossman, M. *The First Nonlinear System of Differential and Integral Calculus.* Rockport, MA: Mathco, 1979.
This book contains a detailed account of the geometric calculus, which was the first of the non-Newtonian calculi. Also included are discussions of the analogy that led to the discovery of that calculus, and some heuristic guides for its application.

Meginniss, J. R. "Non-Newtonian Calculus Applied to Probability, Utility, and Bayesian Analysis." *Proceedings of the American Statistical Association*: Business and Economics Statistics Section (1980), pp. 405-410.
This paper presents a new theory of probability suitable for the analysis of human behavior and decision making. The theory is based on the idea that subjective probability is governed by the laws of a non-Newtonian calculus and one of its corresponding arithmetics.

Grossman, J., Grossman, M., and Katz, R. *The First Systems of Weighted Differential and Integral Calculus.* Rockport, MA: Archimedes Foundation, 1980.
This monograph reveals how weighted averages, Stieltjes integrals, and derivatives of one function with respect to another can be linked to form systems of calculus, which are called weighted calculi because in each such system a weight function plays a central role.

Grossman, J. *Meta-Calculus: Differential and Integral.* Rockport, MA: Archimedes Foundation, 1981.
This monograph contains a development of systems of calculus, called meta-calculi, that transcend the classical calculus, for example in the following manner. In each meta-calculus the gradient, or average rate of change, of a function f on an interval [r,s] depends on *all* the points (x,f(x)) for which $r \leq x \leq s$, whereas the classical gradient $[f(s) - f(r)]/(s - r)$ depends only on the endpoints (r,f(r)) and (s,f(s)). The meta-calculi arose from the problem of measuring stock-price performance when taking all intermediate prices into account.

Grossman, M. *Bigeometric Calculus: A System with a Scale-Free Derivative.* Rockport, MA: Archimedes Foundation, 1983.
This book contains a detailed treatment of the bigeometric calculus, which has a derivative that is scale-free, i.e., invariant under all changes of scales (or units) in function arguments and values. Also included are heuristic guides for the application of that calculus, and various related matters such as the bigeometric method of least squares.

Grossman, J., Grossman, M., and Katz, R. *Averages: A New Approach.* Rockport, MA: Archimedes Foundation, 1983.
This monograph is primarily concerned with a comprehensive family of averages (unweighted and weighted) of functions that arose naturally in the development of non-Newtonian calculus and weighted non-Newtonian calculus. The monograph also contains discussions of some heuristic guides for the appropriate use of averages and an interesting family of means of two positive numbers.

A Natural Interpretation of the ω-Measure of an Interval

In this note we shall provide a natural interpretation of the ω-measure of an interval, which is a fundamental notion in our development of the weighted classical calculus in Chapter 2 of *The First Systems of Weighted Differential and Integral Calculus*.

Let $a_1, \ldots, a_n$ be the n-fold arithmetic partition (page 0) of the interval $[r,s]$, and let s_n represent the unweighted sum

$$(a_2 - a_1) + \cdots + (a_n - a_{n-1}),$$

which equals $s - r$. Then

$$\lim_{n \to \infty} s_n = s - r,$$

which we called the classical extent of $[r,s]$.

Now let t_n represent the WEIGHTED sum

$$\omega(a_1) \cdot (a_2 - a_1) + \cdots + \omega(a_{n-1}) \cdot (a_n - a_{n-1}),$$

where ω is any given classical weight function (page 11). Then

$$\lim_{n \to \infty} t_n = \int_r^s \omega,$$

which therefore may be conceived as a WEIGHTED classical extent of $[r,s]$ and which was called the ω-measure of $[r,s]$ (page 11).

Similarly, a natural interpretation of the ω*-measure of an α-interval (page 32) can be provided.

Jane Grossman
Michael Grossman
Robert Katz

ISBN 0977117014

www.ingramcontent.com/pod-product-compliance
Lightning Source LLC
Chambersburg PA
CBHW032017190326
41520CB00007B/513
9780977117017